GEOGRAPHY NOW!

DESERTS AND POLAR REGIONS AROUND THE WORLD

JEN GREEN

PowerKiDS press
New York

Published in 2009 by The Rosen Publishing Group Inc.
29 East 21st Street, New York, NY 10010

First Edition

Editor: Jon Richards
Designer: Ben Ruocco
Consultant: John Williams

Library of Congress Cataloging-in-Publication Data

Green, Jen.
Deserts and polar regions around the world / Jen Green. – 1st ed.
p. cm. – (Geography now)
Includes index.
ISBN 978-1-4358-2874-2 (library binding)
ISBN 978-1-4358-2960-2 (paperback)
ISBN 978-1-4358-2966-4 (6-pack)
1. Deserts–Juvenile literature. 2. Polar regions–Juvenile literature. I. Title.
QH88.G74 2009
551.41'5–dc22

2008025794

Manufactured in China

Picture acknowledgments:
(t-top, b-bottom, l-left, r-right, c-center)
Front cover Dreamstime.com/Hashim Pudiyapura, 1 istockphoto.com/Kevin Ross, 4-5 Dreamstime.com/ David Morrison, 5c Dreamstime.com/Laurin Rinder, 5br Dreamstime.com/Michael Schofield, 6-7 istockphoto.com/Robert Bremec, 6cl Dreamstime.com/Eric Isselée, 8-9 John McAnulty/Corbis, 8cl Dreamstime.com/Hashim Pudiyapura, 9br Dreamstime.com/Jerry Horn, 10-11 Dreamstime.com/Paul Moore, 10bl Dreamstime.com, 11br istockphoto.com/Kevin Ross, 12-13 istockphoto.com/Kevin Ross, 13cr Dreamstime.com/Kheng Guan Toh, 13br courtesy of United States Department of Energy, 14-15 istockphoto.com/Arno Massee, 15tr istockphoto.com/Jason Hills, 15br Dreamstime.com/Kris Hanke, 16-17 Franck Guiziou/Hemis/Corbis, 16bl Pierre Colombel/Corbis, 17br istockphoto.com/Ellen Ebenau, 18-19 Dreamstime.com/Dmitry Pichugin, 18bl istockphoto.com/Amanda Balmain, 19br Dreamstime.com/Fabrizio Argonauta, 20-21 Dreamstime.com/Ron Sumners, 20bl Takoradee/GNU Free Documentation license, 21 br Paul A. Souders/Corbis, 22-23 Jacques Langevin/Corbis Sygma, 23cl Dreamstime.com/Vasiliy Koval, 23br Louie Psihoyos/Corbis, 24-25 Dreamstime.com/Vito Elefante, 24bl Dreamstime.com/Maxfx, 25br Roger Ressmeyer/Corbis, 26-27 istockphoto.com/Haider Yousuf, 26bl istockphoto.com, 27br Dreamstime.com/Paul Cowan, 28-29 istockphoto.com, 28bl courtesy of NOAA, 29br courtesy of NASA

CONTENTS

What are deserts?

The Earth's driest places are called deserts. Scientists define deserts as places where less than 10 in. (25 cm) of rain falls in a year. Deserts are usually edged by slightly wetter, but still dry, scrublands, called semideserts.

DESERT CONDITIONS

Many deserts lie in the regions close to the equator, called the tropics. Here, it is scorching hot by day, because the cloudless skies provide no shade. However, it gets cold at night, because there are no clouds to prevent the heat from escaping. We usually think of deserts as sandy, but three-quarters of all deserts are rocky or stony.

Life is a constant battle for survival for desert plants and animals. Some living things cannot cope with such harsh conditions.

HARSH HABITAT

Extreme temperatures and a lack of water make deserts very tough places to live, but a surprising number of plants and animals manage to survive there. People live in deserts, too—usually in areas that have natural resources, such as valuable minerals or precious water. However, human use of deserts, such as mining, can harm the environment for other living things.

Death Valley in California (below) is the hottest and driest part of North America. A temperature of 133°F (56°C) was once recorded here.

Atacama Desert

The driest desert on Earth is the Atacama Desert, which runs along the northwest coast of Chile in South America. It has an average rainfall of just 0.03 in. (1 mm) each year, but some parts of it had no rain at all for about 400 years. This very long drought was finally broken when torrential rain fell in 1971.

The barren, moonlike landscape of the Atacama Desert is caused by an extreme lack of water.

Why do deserts form?

Deserts are dry because no rain clouds gather there. Clouds form when warm, damp air rises and cools. Cool air cannot hold as much moisture as warm air, so moisture condenses to form clouds. The lack of rain clouds in deserts occurs for several reasons.

An oasis forms where water seeps to the surface in a desert. Plants draw water from the spring, and people and animals come to drink.

LOCATION OF DESERTS

Many of the world's large deserts lie within two bands, between 15 and 30 degrees north and south of the equator. Warm, dry air from the equator sinks here, creating dry conditions. Some deserts lie far inland, where damp ocean winds cannot reach—but there are also deserts near coasts. Here, cold sea currents cool the air, so any moisture falls before reaching land.

Deserts cover about a quarter of the Earth's surface. Every continent except Europe has deserts. These dunes lie in the Thar Desert in northern India.

OASES

Most deserts do contain water, but it lies deep underground in bands of water-soaked rocks called aquifers. When water trickling downward reaches a layer of rocks through which it cannot pass, it runs sideways, and eventually bubbles to the surface at oases. These green patches are vital for desert people and their animals.

Rain shadows

Dry regions called rain shadows lie on the lee slopes of mountain ranges—slopes facing away from moist ocean winds. As winds blow inland, they rise and cool as they meet the mountains. The moisture condenses to form clouds, which shed their rain before reaching the lee side, where deserts may be found.

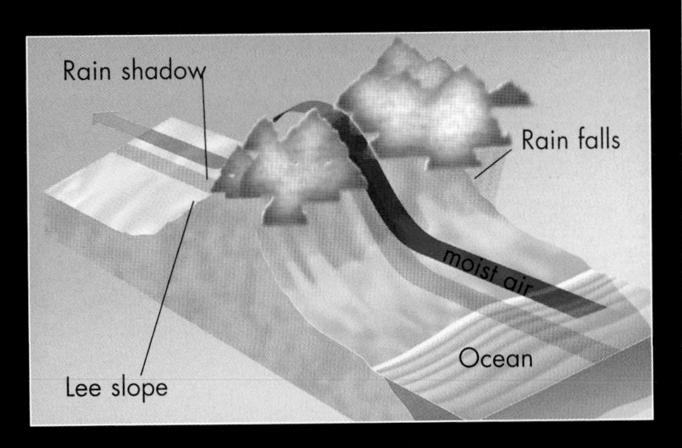

A rain shadow forms where the rocky bulk of a mountain shields an area from damp winds blowing off the ocean.

Shaping deserts

Some deserts contain dramatic features, such as rock pillars, arches, and sand dunes that stretch for hundreds of miles. These have been shaped by wind, ice, and water—a process called erosion.

Large expanses of sand dunes can be found in the deserts of North Africa and Central Asia, and here in the Arabian Desert in the Middle East.

HOW SAND IS FORMED

Extremes of temperature affect rocks in deserts. Rocks expand in the daytime heat and contract in the cold at night. This constant expansion and contraction causes pieces to flake off, creating rocky debris that is carried off by the wind and piles up to form sand dunes. These dunes may be shaped like crescents, stars, or long, straight lines, depending on the direction of the wind.

Wind, ice, and occasional rainstorms have worn down the rock to leave these pinnacles in Monument Valley, in the Navajo Tribal Park on the Utah/Arizona border. Here, there are hundreds of these rocky spires, called hoodoos.

SCOURED BY WIND AND WATER

Gritty sand carried by the wind acts like sandpaper, scouring away at rock surfaces and eating into cracks. In time, this can create unusual features, such as arches and top-heavy pillars. When rain does fall, the desert ground is too dry to absorb it. The water pours into normally dry gullies called wadis, causing flash floods.

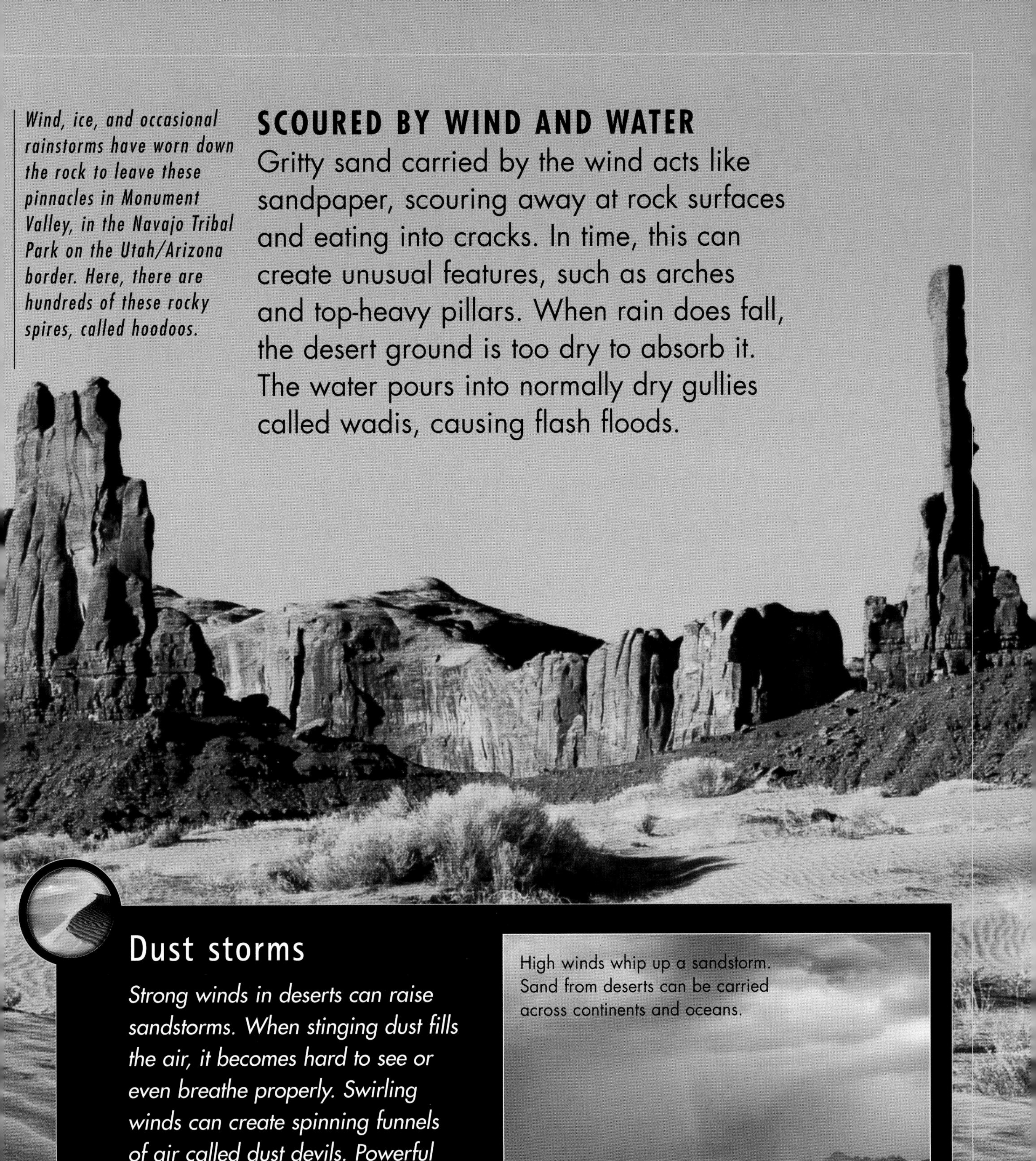

Dust storms

Strong winds in deserts can raise sandstorms. When stinging dust fills the air, it becomes hard to see or even breathe properly. Swirling winds can create spinning funnels of air called dust devils. Powerful winds can carry sand for thousands of miles before dumping it in a distant location.

High winds whip up a sandstorm. Sand from deserts can be carried across continents and oceans.

Desert survival

Plants and animals have to be survival experts to cope with the extreme temperatures and lack of water in deserts. Even so, these harsh habitats are home to plants such as cacti, and animals such as birds and reptiles.

PLANT LIFE

Desert plants, such as the North American creosote bush, have long roots that spread through the ground to suck up every bit of moisture. Plants called succulents store water in their fleshy stems or leaves. Cacti have a thick skin covered with spines. The spines deter animals and also lose less moisture than leaves.

The fennec fox of the Sahara Desert has very large ears that radiate (give out) heat. This helps to keep the animal cool. Its ears are also very good at collecting sounds, helping it to hunt prey at night.

DESERT ANIMALS

Few animals are seen in deserts during the day. Most creatures spend the scorching daylight hours resting in the shade or asleep in cool underground burrows. They emerge to look for food at night, using their senses of smell and hearing to hunt in the darkness. Desert rodents can survive with very little water and live through droughts in a deep sleep similar to hibernation, called estivation.

Saguaro cacti grow in the deserts of the western United States. Their ridged stems can expand to hold water after a rainstorm.

Spadefoot toad

Spadefoot toads of North American deserts spend most of their lives underground. When rain falls, they emerge to lay their eggs in pools of water. Their tadpoles grow up very quickly, becoming adults in just two weeks. These young toads burrow into the mud of the drying pools, and wait for the next rain to fall.

Spadefoot toads have a speeded-up life cycle, so they can grow quickly during any short rainy periods.

Using deserts

Despite the harsh conditions, people have lived in deserts for centuries. Settlements have grown up at oases where crops can be grown, and sometimes close to sources of valuable minerals. About one-twentieth of the world's population lives in deserts.

MINING AND TOURISM

Valuable minerals, such as gold and silver, are sometimes found in deserts. There may also be gemstones, such as diamonds and opals, or rich reserves of oil, coal, and natural gas. Tourism is a growing industry in many desert regions, especially in sandy deserts. People travel out to camps using camels or four-wheel-drive vehicles, to sleep under the stars or to try sports such as sandboarding and rock climbing.

Camel treks are popular with tourists in sandy deserts such as the Sahara. These sand dunes turn rosy-red at dawn and dusk.

CAMPS AND SETTLEMENTS

The land around oases is usually the only part of a desert that can be farmed. Where the land is too dry for crops, people live as nomads—moving from place to place to find grazing and water for livestock such as goats and camels. Even the emptiest deserts are crisscrossed by routes that local people have traveled for centuries to find food or to travel to markets to trade.

Date palms (right) are often planted around oases. As well as dates, these trees yield tough fibers that are used to thatch houses and make ropes and baskets.

Bomb tests

Nuclear weapons tests are sometimes carried out in deserts, because there are no people living there. However, these tests can poison the land for hundreds of years. Nowadays, in places such as the Nevada Desert, these powerful bombs are usually exploded in underground bunkers to minimize any pollution.

The Nevada Desert has been used to test nuclear weapons since the 1950s.

Protecting deserts

You may be surprised to hear that deserts need protection. They are fragile habitats, where plants and animals struggle to survive. When people settle, mine, or try to farm the desert, they bring changes that can easily damage the natural environment.

MINING AND SETTLEMENTS

Some deserts have large mines, where minerals, such as gold, copper, iron, phosphates, or oil, are extracted. Mining damages the environment and can also lead to pollution, such as oil spills. In the United States and Saudi Arabia, deserts are home to big cities such as Las Vegas and Dubai. Here, large numbers of people put a strain on local resources, such as energy, fuel, and water.

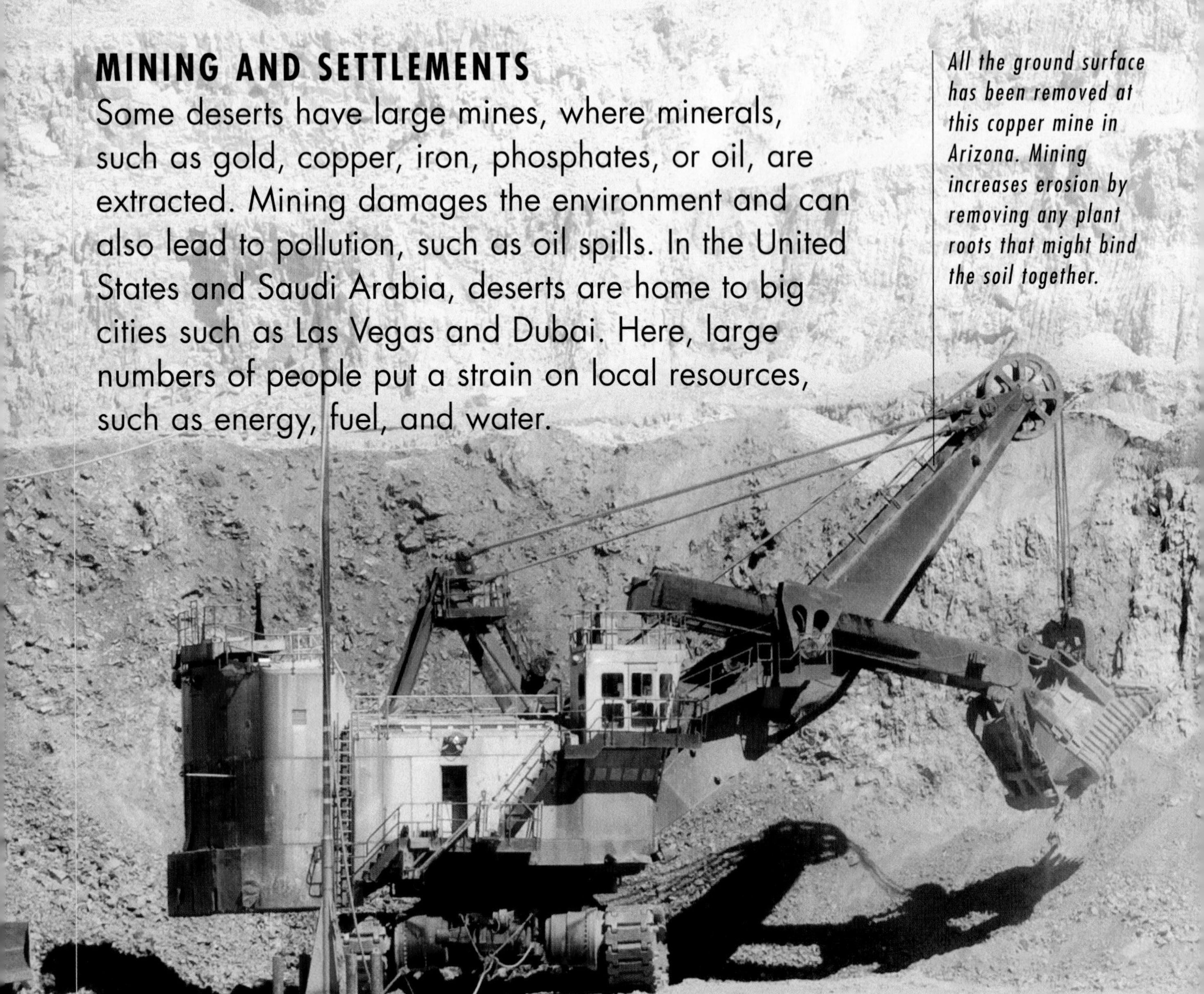

All the ground surface has been removed at this copper mine in Arizona. Mining increases erosion by removing any plant roots that might bind the soil together.

GROWING DESERTS

Many deserts around the world are expanding—a process called desertification. This can happen where people allow livestock, such as goats, to overgraze and eat all the vegetation. More generally, deserts are also spreading because of global warming, which is changing local weather patterns. Some areas of land are becoming drier and turning into deserts.

Goats can quickly destroy trees and shrubs by nipping off greenery and tender shoots.

Irrigation

Deserts can be turned into farmland with the help of irrigation—artificial watering, using sprinklers or other devices. Sprinkler systems use a lot of water and drain local supplies. To preserve these supplies, a special kind of plastic can be added to water so it drains away more slowly, giving plants more time to absorb it.

Rotating sprinklers create circles of green crops in the desert. However, much of the water evaporates before plants have a chance to suck it up.

Largest hot desert

Sahara Desert

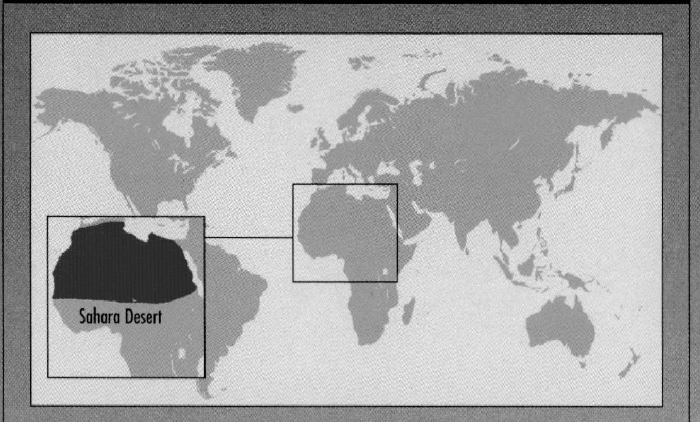

STATISTICS

- *Location: North Africa*
- *Extent: About 3.5 million sq miles (9 million sq km)*
- *Type of desert: Includes sandy, stony, and rocky areas*
- *Largest towns: Timbuktu, Al Azizia, Al Khufra, Siwa*
- *Environmental issues: Desertification (the Sahara is expanding into the Sahel)*

The Sahara is the world's greatest expanse of hot desert, stretching straight across North Africa from the Atlantic Ocean to the Red Sea. About one-fifth of the desert is erg, meaning "seas of sand." The rest consists of stony plains called regs, or rocky areas called hammadas.

EXPANDING DESERT

The Sahara is gradually becoming drier. Rock paintings show that around 6,000 years ago, desert areas were once grassland and home to elephants, hippos, and giraffes. In the last 50 years, the Sahara has expanded into the semidesert region to the south, called the Sahel. This has been partly due to climate change, but also to overgrazing and the cutting of trees for fuel.

This rock painting from Tassili in Algeria shows hippos grazing in the Sahara. It is 5,500 years old and shows that conditions were much wetter at that time.

PEOPLES OF THE SAHARA

The Sahara is very thinly populated. About two million people live here—a tiny number for such an enormous area. Desert people include the Berbers and the Tuareg. The latter are a nomadic people who traditionally live by trading goods, such as salt, and moving their herds in search of pasture. However, many Tuareg have now settled on the edges of towns, where they work in factories or as tourist guides.

The Tuareg wear long robes and headdresses as protection from the burning sun. Camels are the traditional form of transportation.

Sahara town

Timbuktu in Mali is an ancient settlement that grew up where several desert trade routes met. Nomadic peoples, such as the Tuareg, brought salt, dates, and livestock to trade for grain and tea at Timbuktu's markets. Timbuktu is also a center for Islamic learning. Many peoples of the Sahara follow the Muslim faith.

Timbuktu's mosque is an impressive building dating from the fifteenth century. It is made of adobe, which is a mixture of dried mud and straw.

A foggy desert

Namib Desert

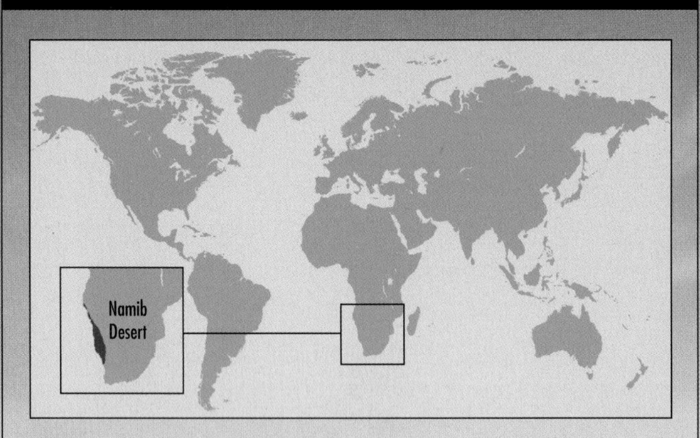

STATISTICS

- *Location: Southwest Africa*
- *Extent: 116,000 sq miles (300,000 sq km)*
- *Type of desert: Includes sand dunes and stony plains*
- *Largest towns: Windhoek, Walvis Bay, Swakopmund*
- *Environmental issues: Pollution and erosion caused by mining on coast*

The Namib Desert runs for 1,180 miles (1,900 km) along the coast of southwest Africa. The only moisture here is provided by fogs that roll in off the Atlantic Ocean. These are created by a cold sea current flowing offshore, which cools the air above it, causing moisture to condense as fog.

Wild elephants wander the barren salt flats of the Etosha Pan in northern Namibia, in search of grazing. This mother and her calf are in the Etosha National Park.

DESERT WILDLIFE

The barren wastes of the Namib are home to small creatures, such as snakes and insects. The sidewinding viper slithers with a sideways motion that keeps most of its body off the burning sand. The darkling beetle does a "handstand" in the early morning to channel dew that has condensed on its body into its mouth. The Namib Desert is also home to large mammals, such as oryx antelopes and even elephants, that migrate between waterholes.

PLANTS OF THE NAMIB

Most plants of the Namib are very short-lived, only sprouting after rain. In contrast, the scraggy-looking welwitschia (below) can live for more than 1,000 years!

Unusual plants of the Namib include the welwitschia. This low-growing plant has long, ragged leaves that gather moisture from fog. Other desert plants appear as if by magic after a shower of rain. They sprout and bloom quickly, turning the desert into a carpet of flowers.

Diamond mining

Diamonds are mined on the coast of the Namib Desert, around the mouth of the Orange River. Large quantities of desert sand are removed to reach the river gravel, which contains the diamonds. Once they have been mined, the rough diamonds are then cut and polished so that they sparkle.

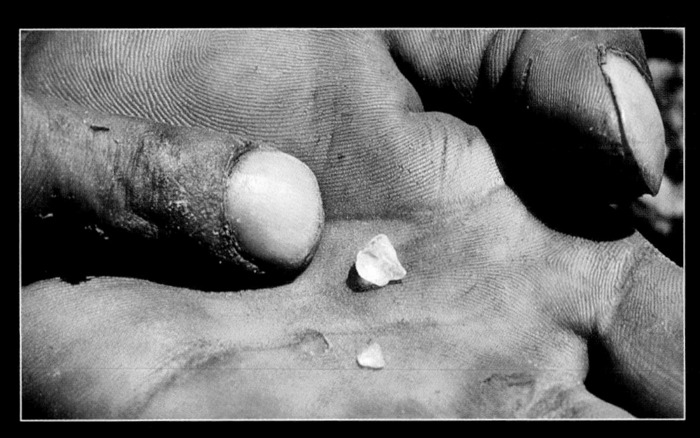

Diamonds may not look like much in their natural state, but they will be worth a lot of money when cut and polished.

Outback life

Australian Outback

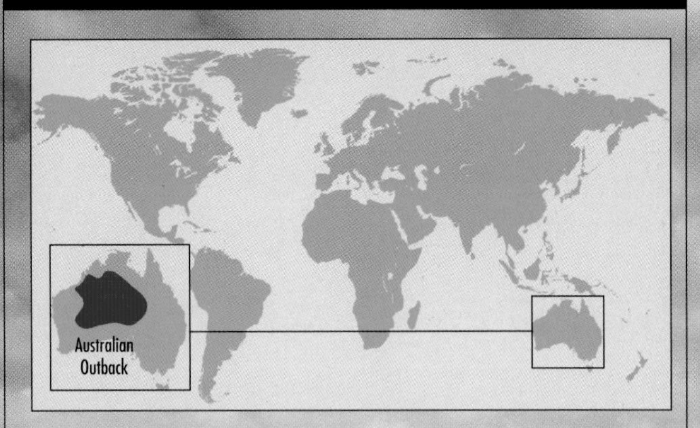

STATISTICS

- Location: Central Australia
- Extent: 599,000 sq miles (1.55 million sq km)
- Type of desert: Sandy or stony plains with large salt lakes
- Largest town: Alice Springs
- Environmental issues: Pollution and erosion caused by mining; fire and desertification caused by drought

Much of inland Australia is covered by four deserts: the Great Sandy, Simpson, Gibson, and Great Victoria Deserts. This region receives slightly more rainfall than some other deserts, allowing dry scrubland called the Outback to form.

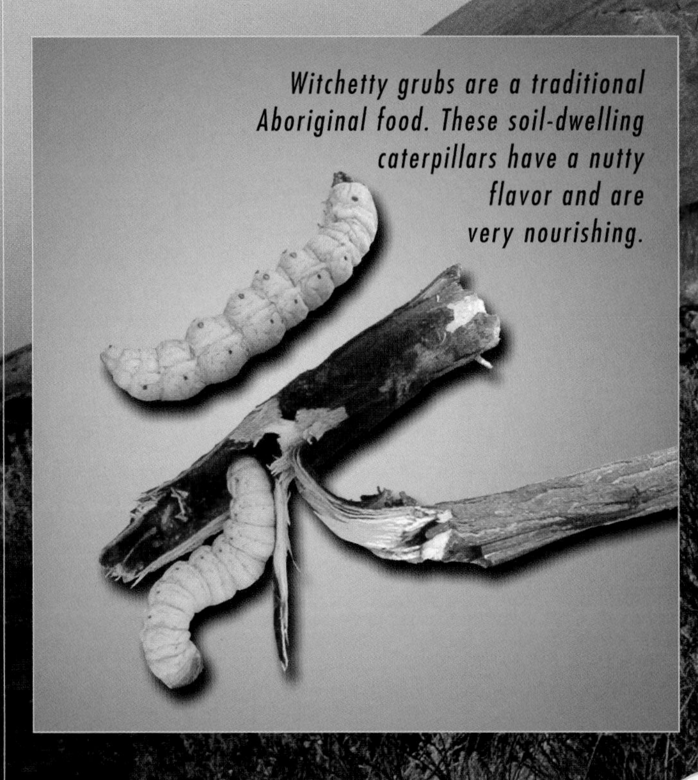

Witchetty grubs are a traditional Aboriginal food. These soil-dwelling caterpillars have a nutty flavor and are very nourishing.

This huge rock outcrop, named Uluru, is sacred to local Aboriginals. Aboriginal culture is celebrated in ceremonies, dance, and song, and also in art.

SALT LAKES AND DROUGHT

The Australian deserts contain huge dry lakes that are covered with salt left by evaporated water. The largest, Lake Eyre, was dry for about a century until rain fell in the 1950s. South of the Great Victoria Desert lies the Nullarbor Plain, whose name means "no tree." Australia has suffered bad droughts in recent years, with devastating fires sparked by parched conditions.

ABORIGINALS

Australian Aboriginals have lived in the Outback for over 40,000 years. They traditionally lived as nomads, hunting animals such as kangaroos, and gathering local plants and small animals to eat. From the 1800s, Europeans took over Aboriginal lands for farming and mining. Forced to give up their nomadic life, most Aboriginals now live in towns.

Opal mining center

Australian deserts are rich in minerals including gold, uranium, and a type of precious stone called opal. The opal mining center of Coober Pedy lies in a very hot desert in the middle of Australia. Many miners' homes have been built into underground caves and tunnels, which remain cool in the stifling heat.

This underground cave home at Coober Pedy has been hollowed out of rock.

Wandering the wastes

Gobi Desert

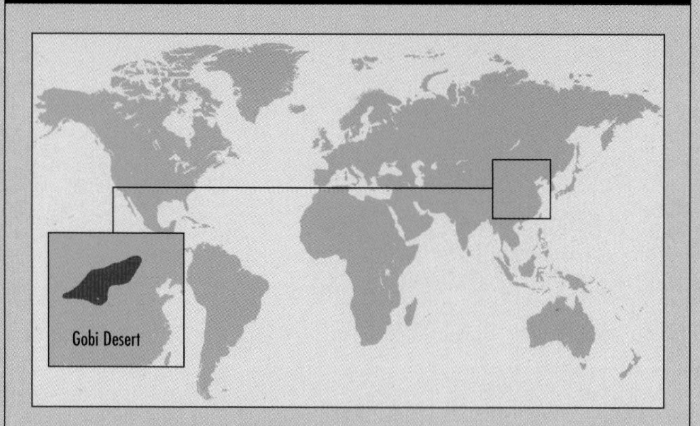

STATISTICS

- *Location: Central Asia*
- *Extent: 402,000 sq miles (1.04 million sq km)*
- *Type of desert: High rocky and sandy plateau*
- *Largest towns: Hohhot, Erenhot, Ulan Bator*
- *Environmental issues: Some desert animals endangered by hunting*

The Gobi Desert in Central Asia is located on a high plateau over 3,280 ft. (1,000 m) above sea level. Temperatures rise to 86°F (30°C) in the summer, but drop to -22°F (-30°C) in the winter, making it the coldest desert outside of Antarctica.

BLEAK CONDITIONS

The Gobi lies between Mongolia and China. To the north, it merges with high, windswept grasslands called the steppes. To the west lies another incredibly bleak desert, the Takla Maklan, whose name means "go in, and you will never come out." Parts of the Gobi and Takla Maklan are sandy desert, with thousands of crescent-shaped dunes called barchan dunes. These form when winds blow mainly from one direction.

NOMADIC HERDERS

The Gobi is inhabited by Mongolian herders, who wander these barren lands with their flocks of sheep and goats, and camp wherever they find grazing. These herders are expert horsemen, who also ride shaggy Bactrian camels. Some former herders have moved to nearby towns, but many still live a nomadic lifestyle. Some now use motorcycles to round up their flocks.

A Mongolian herder on a camel drives his goats. Bactrian camels have two humps, unlike the Arabian camel, or dromedary, which only has one hump. The Gobi species of Bactrian camel has an extra-thick winter coat.

Mongolian tents, called yurts, consist of felt and canvas stretched over a wooden framework. These homes are light to carry and quick to put up.

Dinosaur eggs

In 1922, a U.S. scientist, Roy Chapman Andrews, discovered fossilized dinosaur eggs in the Gobi Desert. This was an important find, because it was the first evidence that dinosaurs laid eggs, just like modern reptiles. Hundreds more dinosaur fossils have since been found in the Gobi Desert, where they have been preserved by the desert sand.

This fossilized dinosaur egg (below) contains the remains of an oviraptor, and on the left is the skull of a baby dromeosaur.

Desert cities

Mojave and Sonoran

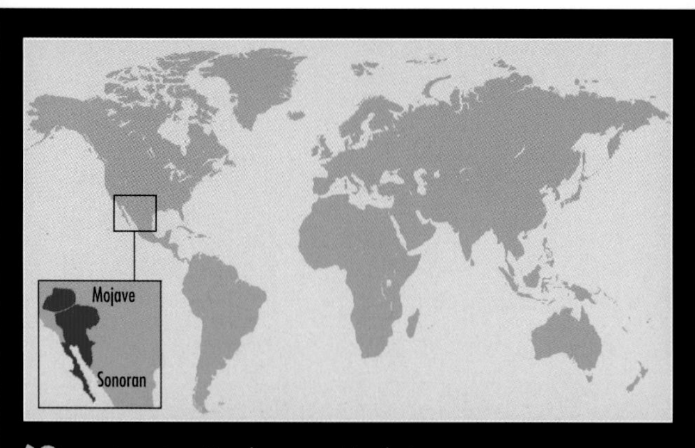

STATISTICS

- *Location: Southwestern North America*
- *Extent: Mojave—25,000 sq miles (64,000 sq km); Sonoran—120,000 sq miles (310,800 sq km)*
- *Type of desert: Includes stony, rocky, and sandy terrain*
- *Largest cities: Phoenix, Tucson, Las Vegas*
- *Environmental issues: Low water supplies and pollution*

Much of the southwestern United States is covered by deserts, such as the Sonoran and Mojave Deserts. Somewhat surprisingly, this very dry region has many towns and several major cities, such as Phoenix, Arizona, and Las Vegas, Nevada.

Las Vegas receives just 4 in. (10 cm) of rain a year. This golf course is kept lush with water from local reservoirs and aquifers. This means there is less water for farming.

FIRST SETTLEMENTS

The very dry conditions in the Mojave are partly caused by nearby mountains, which block moist ocean winds. The original inhabitants of these deserts were Native American groups such as the Papago, who farmed using irrigation. Early desert homes were a labyrinth of caves built in the cliffs to provide shade from the sun.

LAS VEGAS

The modern city of Las Vegas grew quickly as a center for gambling. Las Vegas draws on distant water sources to irrigate its parks and golf courses. This depletes the region's supplies. Huge amounts of energy are needed to power the air-conditioning units and other appliances that make life comfortable in this hot desert city.

Bright lights and air-conditioning units in the high-rise, luxury casinos of Las Vegas use a lot of energy.

Desert energy

Some of Las Vegas' energy comes from power stations that burn coal and oil. These contribute to the air pollution that is causing global warming. An ideal alternative source of energy comes from solar power stations, which harness the sun's heat and light to create electricity without causing pollution.

The Mojave solar power station is currently the largest of its kind in the world.

Oil-rich desert

The Arabian Desert covers a large part of five countries: Saudi Arabia, Oman, Yemen, Jordan, and the United Arab Emirates (UAE). The discovery of oil has brought wealth and great change to this region.

Arabian Desert

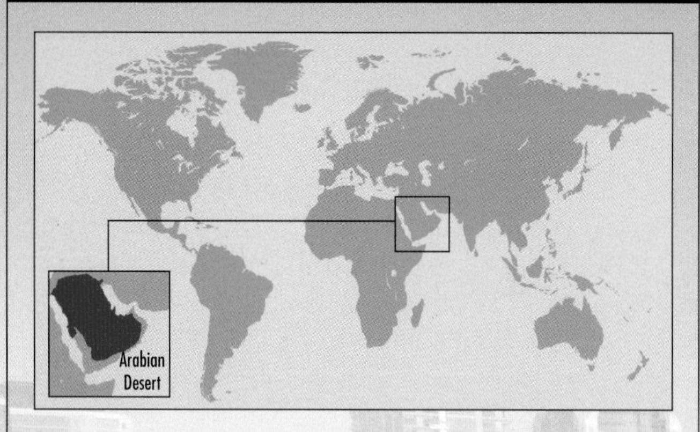

STATISTICS

- *Location: Arabian Peninsula, southwest Asia*
- *Extent: 501,932 sq miles (1.3 million sq km)*
- *Type of desert: Sand-seas, also stony and rocky terrain*
- *Largest cities: Riyadh, Abu Dhabi, Dubai, Aden, Muscat*
- *Environmental issues: Pollution from oil industry, and cities depleting water supply*

Oil from desert wells is piped to refineries for processing. The Trans-Arabian Pipeline is one of the world's longest, running for 1,060 miles (1,700 km), although it is no longer in use.

EMPTY QUARTER

The Arabian Desert includes one of the world's largest sand-seas, the Rub'al Khali or Empty Quarter. The dunes here include long seif (or sword) dunes. For centuries, this vast wilderness was known only to Bedouin tribes, who traditionally lived as nomadic herders and traders.

Cities such as Dubai, shown here, have gleaming skyscrapers, palaces, and fine mosques built with oil money. The region produces almost a quarter of the world's oil.

WEALTH FROM OIL

Oil was discovered in the Arabian Desert in the 1930s. Modernization followed and the desert is now crisscrossed by roads used by four-wheel-drive vehicles instead of camels. One aspect of life is still traditional—people follow the Muslim faith, which began here nearly 1,500 years ago. Muslim pilgrims journey to the holy city of Mecca in the western desert.

Desalination

Cities on the Arabian coast get much of their fresh water by removing salt from seawater. This process, called desalination, is expensive, so only wealthy countries can afford it. Desalination also provides some water for irrigation schemes that are greening parts of the desert.

Desalination plants like this provide Arabian cities with water for drinking and irrigating city parks.

Frozen desert

Antarctica

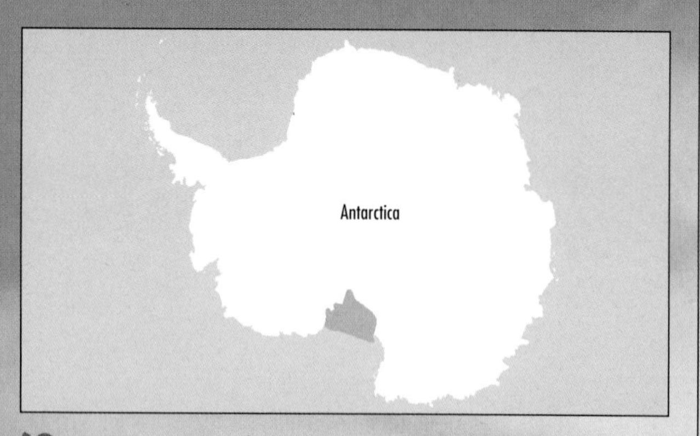

STATISTICS

- *Location: Southern Ocean*
- *Extent: 5.4 million sq miles (13.9 million sq km)*
- *Type of desert: Covered by thick ice cap*
- *Largest settlement: McMurdo Base, about 1,000 scientists*
- *Environmental issues: Coastal ice is melting due to global warming*

Where in the world can you find a desert covered with frozen water? The answer is Antarctica, which is a desert because so little rain falls there. It hardly ever rains because in the very cold climate, moisture cannot evaporate to form clouds.

These scientists are taking part in a study of Weddell seals. The climate is too harsh inland for wildlife to survive, but animals such as seals and penguins are plentiful on the coasts.

ICE CAP

Antarctica is the coldest place on Earth. The continent is covered by a thick cap of ice that is up to 2.5 miles (4 km) deep. Antarctica has not always been ice-covered, however—scientists have found fossilized plants and animals that show it was once much warmer.

SCIENTIFIC RESEARCH

Antarctica is the only desert on Earth that has never been inhabited. Under the terms of an international treaty, it belongs to no country and is used only for scientific research. Up to 4,000 scientists stay here in research bases, mainly in the summer, to study the continent's ice, oceans, rocks, climate, and wildlife.

Antarctica has high mountains rising to over 16,400 ft. (5,000 m). Scientists have discovered that Antarctica is made up of several large landmasses and many islands that are buried beneath the ice.

Ozone hole

Antarctic scientists have made several major discoveries. In 1985, British scientists discovered that the ozone layer in the atmosphere above Antarctica had become much thinner. Chemicals called CFCs were to blame. These have now been banned to give the ozone layer a chance to recover.

This false-color image taken by a satellite in 2001 shows the ozone hole over Antarctica, colored blue.

Glossary, Further Information, and Web Sites

ADOBE
Bricks made from mud and straw.

AQUIFER
A layer of water-soaked rocks underground.

BARCHAN DUNE
A crescent-shaped sand dune.

CFCs (CHLOROFLUOROCARBONS)
Chemicals that were used to make aerosol sprays and refrigerators. CFCs are now banned, because they damage the Earth's protective ozone layer.

CONDENSE
When water changes from gas to liquid.

DESALINATION
The process of making fresh water by removing salt from seawater.

DESERTIFICATION
When a desert expands into the dry scrublands that surround it.

DUST DEVIL
A whirling column of air that picks up sand in a desert.

ERG
An expanse of sandy desert.

EROSION
When rock or soil is worn away and carried off by water, ice, or the wind.

ESTIVATE
When an animal enters a deep sleep similar to hibernation, in order to survive heat or drought.

EVAPORATE
When water changes from a liquid into a gas.

FLASH FLOOD
When water rises quickly in a stream or river after heavy rain.

GLOBAL WARMING
An increase in temperatures around the world. Many scientists are now convinced that global warming is caused by air pollution.

HAMMADA
A rocky desert.

HOODOO
A top-heavy pillar of rock shaped by the wind.

IRRIGATE
When farmers water the land in order to grow crops.

LEE SLOPE
The side of a mountain that faces away from the prevailing wind.

MIGRATE
When animals move on seasonal journeys, usually to their regular feeding grounds.

NOMADS
A group of people who move from place to place, often to find grazing or water for their animals.

OASIS
A place where water reaches the surface in a desert.

PLATEAU
A large, flat area of land that is raised above the surrounding countryside.

RAINSHADOW
A dry area that forms where a mountain shields the land from moist winds.

REG
A stony desert.

SUCCULENT
A plant, such as a cactus, that stores water inside thick fleshy stems or leaves.

WADI
A dry gully in a desert that fills with water after rain.

WEATHERING
When forces, such as water and wind, wear away the rocks on Earth's surface.

FURTHER READING

Deserts: Surviving in the Sahara
by Michael Sandler
(Bearport Publishing, 2005)

Earth Files: Deserts
by Anita Ganeri
(Gareth Stevens, 2002)

The World's Top Ten: Deserts
by Neil Morris
(Raintree Steck-Vaughn, 1997)

WEB SITES

Due to the changing nature of Internet links, PowerKids Press has developed an online list of Web sites related to the subject of this book. This site is updated regularly. Please use this link to access this list:
www.powerkidslinks.com/geon/despolar

Deserts topic web

Use this topic web to discover themes and ideas in subjects that are related to deserts.

DESERTS

GEOGRAPHY

- How deserts form, and how climate and location are involved in their creation.
- How weathering and erosion shape deserts and features within them, such as sand dunes.
- Natural resources in deserts, such as minerals and gemstones.
- Energy from the desert, including oil reserves and solar power.

SCIENCE AND THE ENVIRONMENT

- Desert wildlife and how plants and animals have adapted to harsh desert conditions.
- Environmental problems in the desert, such as industry and pollution.
- How varying conditions can affect the rate of desertification.
- Conservation work to tackle many of these environmental problems.

ART AND CULTURE

- How people live and survive in the desert—living as nomads or settling near oases.
- Development of traditional clothing and how it suits desert conditions.
- The art and culture of desert people, such as cave paintings, jewelry, and religion.

ENGLISH AND LITERACY

- Stories and accounts of the lives of desert peoples, such as nomadic tribes, and those who work in deserts, such as oil workers and scientists.
- Debate the pros and cons of development in deserts—for example, mining and desert irrigation for farming.

HISTORY AND ECONOMICS

- Use of desert resources such as minerals.
- Development of desert towns and industries.
- The growth of trade routes through deserts, and how these have led to the development of market towns along those routes.

Index